Beyond the Stars: Fascinating Facts about the Universe

Index

I. Chapter 1: The Vastness of Space
A. The scale of the universe: from galaxies to superclusters
B. Mind-boggling distances and measurements
C. The concept of cosmic expansion and the Big Bang Theory

II. Chapter 2: Celestial Bodies
A. The majestic stars: types, life cycles, and sizes
B. Exploring planets and their unique characteristics
C. Moons, asteroids, and comets: fascinating companions of our solar system

III. Chapter 3: Stellar Phenomena
A. Supernovae: explosive stellar deaths that shape the cosmos
B. Neutron stars and pulsars: the remnants of massive stars
C. Black holes: the enigmatic gravitational powerhouses

IV. Chapter 4: The Mysteries of Dark Matter and Dark Energy
A. Understanding the elusive nature of dark matter
B. Dark energy and its role in the accelerating expansion of the universe
C. Current research and theories surrounding these cosmic enigmas

V. Chapter 5: Exoplanets and the Search for Life
A. The discovery of exoplanets and their potential for habitability
B. The characteristics of potentially habitable exoplanets
C. The ongoing search for extraterrestrial life and the implications for humanity

VI. Chapter 6: Cosmic Events and Phenomena
A. Eclipses, meteor showers, and other celestial events
B. Gamma-ray bursts and their energetic impact
C. Gravitational waves: ripples in the fabric of spacetime

VII. Chapter 7: Space Exploration and Discoveries
A. Milestones in space exploration: from the Moon landing to Mars missions
B. Robotic probes and their contributions to our understanding of the universe
C. The future of space exploration and the prospects for interstellar travel

VII. Chapter 8: The Human Connection to the Universe
A. Cultural and historical perspectives on the cosmos
B. The role of astronomy in shaping our understanding of the world
C. Reflections on our place in the vastness of the universe

IX. Chapter 9: Bonus Facts about Space

I. Chapter 1: The Vastness of Space

A. The scale of the universe: from galaxies to superclusters

- Galaxies, such as our Milky Way, are vast systems containing billions or even trillions of stars, along with various celestial objects like planets, asteroids, and comets.

- The observable universe is estimated to contain over 100 billion galaxies, each with its unique collection of stars and celestial bodies.

- The distance between galaxies can be enormous. On average, galaxies are separated by millions of light-years, with each light-year representing the distance light travels in one year, roughly 5.88 trillion miles (9.46 trillion kilometres).

- Superclusters are enormous structures that contain multiple galaxy clusters, forming some of the largest known cosmic structures in the universe.

- The Great Wall, a massive supercluster, is one of the largest known structures in the universe. It stretches over 500 million light-years and contains thousands of galaxies.

- At the largest scales, the universe exhibits a web-like structure known as the cosmic web. It consists of vast voids, filaments, and sheets, with galaxies and superclusters arranged along these structures.

- The Local Group is a small cluster of galaxies that includes the Milky Way, Andromeda, and about 54 other smaller galaxies. It spans about 10 million light-years in diameter.

- The largest known galaxy cluster is the Hercules-Corona Borealis Great Wall, stretching over 10 billion light-years across.

- The scale of the universe is constantly expanding due to the phenomenon known as cosmic expansion. This means that galaxies and superclusters are gradually moving away from one another over time.

- Despite the immense size of the universe, our understanding and exploration of it are still limited. The true extent of the cosmos is yet to be fully comprehended, leaving room for continued discoveries and revelations.

B. Mind-boggling distances and measurements

- The speed of light, the fastest known speed in the universe, is approximately 299,792 kilometres per second (186,282 miles per second). It can travel around the Earth's equator about 7.5 times in just one second.

- The nearest star system to our solar system, Alpha Centauri, is about 4.37 light-years away. This means that the light we see from Alpha Centauri today left the star over four years ago.

- The Milky Way galaxy, which is our home galaxy, has a diameter of about 100,000 light-years. It would take light 100,000 years to travel from one end of the galaxy to the other.

- The Andromeda Galaxy, the closest spiral galaxy to the Milky Way, is approximately 2.537 million light-years away. It is so distant that the light we observe from it today began its journey when early humans were just beginning to use tools.

- The distance from Earth to the edge of the observable universe is estimated to be about 46.5 billion light-years. This is due to the expansion of space over time.

- The Voyager 1 spacecraft, launched in 1977, is currently the farthest human-made object from Earth. As of 2021, it has travelled over 22 billion kilometres (13.7 billion miles) and is still sending valuable data about interstellar space.

- The Oort Cloud, a hypothetical region of icy objects beyond the outer reaches of our solar system, is estimated to extend about 0.8 light-years from the Sun. It is believed to be the source of long-period comets.

- The distance between the Earth and the Sun, known as an astronomical unit (AU), is approximately 149.6 million kilometres (93 million miles). It takes light about 8 minutes and 20 seconds to travel this distance.

- In terms of mass, the Sun is about 333,000 times heavier than Earth. It accounts for about 99.86% of the total mass of the entire solar system.

- The largest known star, UY Scuti, is a red supergiant located around 9,500 light-years away from Earth. It has a diameter estimated to be about 1,700 times that of the Sun, which means it could engulf the entire orbit of Jupiter.

C. The concept of cosmic expansion and the Big Bang Theory

- The Big Bang Theory proposes that the universe originated from an incredibly hot and dense state approximately 13.8 billion years ago. It suggests that all matter and energy were concentrated in an infinitesimally small point before rapidly expanding and giving rise to the universe as we know it.

- The evidence supporting the Big Bang Theory includes the observation of the cosmic microwave background radiation, which is the residual heat left over from the early stages of the universe. This radiation is detected uniformly in all directions and provides strong evidence for the expansion of the universe.

- The concept of cosmic expansion is based on the observation that galaxies are moving away from one another. The farther a galaxy is from us, the faster it appears to be receding. This observation led to the formulation of Hubble's Law, which states that the velocity at which a galaxy moves away from us is proportional to its distance.

- Edwin Hubble, an American astronomer, made the groundbreaking discovery in the 1920s that galaxies beyond our Milky Way were receding from us, providing evidence for the expansion of the universe. This observation has since been confirmed and refined through various methods.

- The expansion of the universe is not a movement of galaxies through space but rather the stretching of space itself. It is often analogized to the inflation of a balloon, where dots on the surface move farther apart as the balloon expands.

- The rate of cosmic expansion is described by the Hubble constant, which quantifies the current rate at which the universe is expanding. Determining an accurate value for the Hubble constant is an ongoing challenge in cosmology and has implications for understanding the age and future of the universe.

- In the early stages of the universe's expansion, a rapid phase called cosmic inflation is hypothesized to have occurred. This inflationary period would have led to the uniformity of the cosmic microwave background radiation observed today and set the stage for the formation of galaxies and large-scale structures.

- While the Big Bang Theory describes the origin and early evolution of the universe, it does not provide a definitive explanation for what triggered the initial expansion or what conditions were present before the Big Bang. These are still active areas of research and debate in cosmology.

- The concept of cosmic expansion implies that the universe has been continuously expanding since the Big Bang. However, it is important to note that expansion does not imply that the universe is expanding into something or from a specific centre.

- The discovery of cosmic expansion and the formulation of the Big Bang Theory revolutionized our understanding of the universe and laid the foundation for modern cosmology. It has provided a framework for exploring the evolution, structure, and ultimate fate of the cosmos.

II. Chapter 2: Celestial Bodies

A. The Majestic Stars: Types, Life Cycles, and Sizes

- Stars come in various types, including red dwarfs, yellow dwarfs (like our Sun), blue giants, and red supergiants, each distinguished by colour, temperature, and size.

- The largest known star, UY Scuti, is a red supergiant with a diameter estimated to be about 1,700 times that of the Sun. Its immense size could engulf the entire orbit of Jupiter.

- Stellar life cycles vary depending on their mass. Low-mass stars, like red dwarfs, can burn for trillions of years, while massive stars may only last a few million years before undergoing a supernova explosion.

- During a supernova, a massive star explodes, releasing an enormous amount of energy and scattering heavy elements into space. These explosions can briefly outshine entire galaxies.

- Neutron stars are incredibly dense stellar remnants that form when a massive star undergoes a supernova. They can have a mass greater than the Sun but are squeezed into a size similar to a city, resulting in extreme gravitational forces.

- Black holes are formed when the core of a massive star collapses under its gravity. They possess such strong gravitational pull that nothing, not even light, can escape their grasp beyond the event horizon.

- White dwarfs are remnants of low-mass stars that have exhausted their nuclear fuel. They are incredibly dense and hot, representing the final stage of stellar evolution for stars like our Sun.

- Red giants are evolved stars that have exhausted their core hydrogen fuel and expanded in size. They can be hundreds of times larger than their original size and are characterized by their reddish colour.

- Variable stars, such as Cepheid variables, pulsate in size and brightness over time. This property allows astronomers to determine their distance by analyzing their pulsation patterns.

- Binary star systems consist of two stars gravitationally bound to each other. They can have various configurations, including close binaries where the stars orbit closely, or wide binaries where they are separated by a significant distance.

B. Exploring Planets and Their Unique Characteristics

- Mercury is the smallest planet in our solar system and is closest to the Sun. It experiences extreme temperature variations, with scorching hot days and cold nights.

- Venus, often called Earth's "sister planet," has a thick atmosphere composed mainly of carbon dioxide, causing a runaway greenhouse effect. It is the hottest planet in our solar system, with surface temperatures reaching over 450 degrees Celsius (842 degrees Fahrenheit).

- Earth is the only known planet to support life. It has a diverse range of ecosystems, an oxygen-rich atmosphere, and liquid water, making it a unique and habitable world.

- Mars, often referred to as the "Red Planet," has intrigued scientists due to its potential for past or present microbial life. It has distinctive features like polar ice caps, vast deserts, and the largest volcano in the solar system, Olympus Mons.

- Jupiter is the largest planet in our solar system, with a diameter over 11 times that of Earth. It is famous for its swirling cloud bands, the Great Red Spot (a giant storm), and its vast system of moons.

- Saturn is known for its magnificent ring system, composed of countless particles of ice and rock. It is the second-largest planet and has a distinctive yellowish colour due to its atmospheric composition.

- Uranus is an ice-giant planet tilted on its side, making it unique in our solar system. It has a predominantly hydrogen and helium atmosphere, and its bluish colour is a result of methane gas in its upper atmosphere.

- Neptune, the farthest known planet from the Sun, is also an ice giant. It exhibits strong winds and features a deep blue colouration due to the presence of methane in its atmosphere.

- Gas giants like Jupiter and Saturn lack a solid surface and are primarily composed of hydrogen and helium. They have turbulent atmospheres and are known for their impressive storms and giant swirling vortices.

- Exoplanets are planets that orbit stars outside our solar system. Their discovery has expanded our understanding of planetary systems and raised the possibility of finding habitable worlds beyond Earth.

C. Moons, Asteroids, and Comets: Fascinating Companions of Our Solar System

- Earth's Moon is the fifth-largest moon in the solar system and the only one humans have visited. It plays a crucial role in stabilizing Earth's rotation and tides.

- Io, one of Jupiter's four largest moons, is the most volcanically active object in our solar system. It experiences intense tidal forces from Jupiter, causing frequent volcanic eruptions.

- Titan, Saturn's largest moon, has a thick atmosphere and hydrocarbon lakes. It is the only known moon to have a substantial atmosphere and is of great interest in the search for extraterrestrial life.

- Phobos and Deimos are the two moons of Mars. They are irregularly shaped and are likely captured asteroids rather than formed alongside Mars.

- Ganymede, the largest moon in the solar system, is one of Jupiter's moons and is even larger than the planet Mercury. It has a magnetic field and is geologically diverse.

- Enceladus, one of Saturn's moons, has a subsurface ocean beneath its icy crust. It is known for its geysers that spew water vapour and icy particles into space, indicating the potential for liquid water and potentially habitable conditions.

- Asteroids are rocky objects that orbit the Sun, primarily located in the asteroid belt between Mars and Jupiter. They vary in size, with the largest asteroid, Ceres, being classified as a dwarf planet.

- The Kuiper Belt is a region beyond Neptune's orbit that contains numerous icy bodies, including Pluto, which was reclassified as a dwarf planet in 2006. It is believed to be a remnant of the early solar system.

- Comets are icy bodies that originate from the outer regions of the solar system. As they approach the Sun, they develop a glowing coma and, sometimes, a visible tail due to the heating and outgassing of volatile substances.

- The famous Halley's Comet, visible from Earth approximately once every 76 years, is perhaps the most well-known comet. It last appeared in 1986 and will be visible again in 2061.

- These facts shed light on the diverse and captivating aspects of stars, planets, moons, asteroids, and comets, enriching our understanding of the fascinating companions that populate our solar system.

III. Chapter 3: Stellar Phenomena

A. Supernovae: Explosive Stellar Deaths that Shape the Cosmos

- Supernovae are powerful and catastrophic events that occur when massive stars reach the end of their lives and explode in a brilliant display of light and energy.

- There are two main types of supernovae: Type I and Type II. Type I supernovae result from the explosive fusion of a white dwarf in a binary star system, while Type II supernovae occur when a massive star runs out of fuel and undergoes a gravitational collapse.

- The explosion of a supernova releases an enormous amount of energy, briefly outshining entire galaxies and emitting intense radiation across the electromagnetic spectrum.

- Supernovae play a crucial role in the creation of heavy elements. During the explosion, nuclear fusion occurs, producing elements like iron, gold, and uranium, which are scattered into space and eventually incorporated into new stars and planetary systems.

- The remnants of a supernova explosion form a supernova remnant, which includes an expanding shell of gas and dust. These remnants can persist for thousands of years and provide fertile grounds for the formation of new stars.

- The Crab Nebula, located approximately 6,500 light-years away from Earth, is one of the most famous supernova remnants. It formed from a supernova observed by astronomers in the year 1054.

- Supernovae are rare events in our galaxy, with an estimated frequency of around 1 to 3 per century. However, they are more common in other galaxies, where numerous supernovae can be observed each year.

- The energy released during a supernova explosion is so immense that the supernova can outshine its entire host galaxy for a brief period. This allows astronomers to detect supernovae in distant galaxies and study their properties.

- Supernovae are classified based on their light curves and spectra. The study of supernovae has led to the discovery of different subclasses, including Type Ia, Type Ib, Type Ic, and Type II-P.

- Supernovae have a profound impact on the evolution and dynamics of galaxies. The energy and shockwaves they generate can trigger the formation of new stars, enrich interstellar gas with heavy elements, and shape the structure of galaxies over cosmic timescales.

B. Neutron Stars and Pulsars: The Remnants of Massive Stars

- Neutron stars are incredibly dense remnants that form when the core of a massive star collapses under its gravity after a supernova explosion.
- A teaspoon of neutron star material would weigh billions of tons on Earth, making neutron stars one of the densest known objects in the universe.
- Neutron stars are typically only about 20 kilometres (12 miles) in diameter, but they can have masses several times that of the Sun, resulting in extreme gravitational forces.
- Pulsars are a type of neutron star that emit beams of electromagnetic radiation, often in the form of radio waves, from their magnetic poles. As the star rotates, these beams sweep across space like a lighthouse, causing periodic pulses to be observed on Earth.
- The first pulsar, PSR B1919+21, was discovered in 1967 by Jocelyn Bell Burnell and Antony Hewish. Its regular pulsing pattern initially puzzled astronomers and led to its nickname "LGM-1" (Little Green Men-1).
- Pulsars can rotate at incredibly high speeds, ranging from milliseconds to several seconds per rotation. The fastest known pulsar, PSR J1748-2446ad, completes a rotation in just 1.4 milliseconds.
- Neutron stars exhibit strong magnetic fields, which can be a trillion times stronger than Earth's magnetic field. These intense magnetic fields play a significant role in shaping the behaviour and emissions of neutron stars.
- Some neutron stars are observed to emit powerful beams of X-rays, known as X-ray pulsars. These pulsars are thought to be accreting matter from a nearby companion star, generating intense X-ray emissions as the matter falls onto the neutron star's surface.
- Neutron stars can also undergo intense bursts of X-rays and gamma rays known as X-ray bursts. These bursts result from nuclear reactions and instabilities occurring on the surface of the neutron star.
- Magnetars are a rare type of neutron star with extremely powerful magnetic fields. They exhibit sporadic bursts of X-rays and gamma rays and are thought to be associated with highly magnetized regions on the star's surface.

C. Black Holes: The Enigmatic Gravitational Powerhouses

- Black holes are regions in space where gravity is so strong that nothing, not even light, can escape their gravitational pull. They are formed from the remnants of massive stars that undergo gravitational collapse.

- The boundary of a black hole, beyond which nothing can escape, is called the event horizon. It marks the point of no return for anything crossing it.

- The Schwarzschild radius defines the size of a non-rotating black hole. It represents the distance from the singularity to the event horizon and determines the size of the black hole's event horizon.

- Black holes have three fundamental properties: mass, spin (angular momentum), and charge. The "no-hair theorem" states that these properties uniquely define a black hole, regardless of the objects that formed it.

- Supermassive black holes are millions or even billions of times more massive than the Sun. They reside at the centres of most galaxies, including our own Milky Way, and play a crucial role in galactic evolution.

- Black holes can grow in mass by accreting matter from their surroundings. As matter falls toward a black hole, it forms an accretion disk that emits intense radiation across the electromagnetic spectrum.

- The first-ever direct image of a black hole's event horizon was captured by the Event Horizon Telescope in 2019. The image revealed the silhouette of a supermassive black hole at the centre of the galaxy M87.

- Black holes can emit powerful jets of energetic particles and radiation from their poles. These jets can extend for thousands of light-years and have a significant impact on the surrounding galactic environment.

- Black holes can influence the motion of nearby stars and gas clouds through their immense gravitational pull. This effect has been observed in binary star systems, where a black hole's presence can cause detectable changes in the orbits of companion stars.

- The study of black holes has provided valuable insights into the nature of gravity, the curvature of spacetime, and the extreme conditions that exist within these enigmatic objects. Black holes continue to be a subject of intense research and exploration in astrophysics.

IV. Chapter 4: The Mysteries of Dark Matter and Dark Energy

A. Understanding the Elusive Nature of Dark Matter

- Dark matter is a mysterious form of matter that does not interact with light or other forms of electromagnetic radiation, making it invisible to direct observation.

- It is estimated that dark matter constitutes about 85% of the total matter in the universe, while ordinary matter (atoms) accounts for only around 5%.

- The presence of dark matter is inferred from its gravitational effects on visible matter and the structure of the universe, such as the rotation curves of galaxies and the distribution of matter in galaxy clusters.

- Despite extensive efforts, scientists have yet to directly detect or identify dark matter particles. Various candidates, such as WIMPs (Weakly Interacting Massive Particles), axions, and sterile neutrinos, have been proposed, but their existence remains unconfirmed.

- The nature of dark matter remains one of the most significant mysteries in astrophysics and particle physics. Understanding its properties could revolutionize our understanding of the universe and the fundamental laws of physics.

- Dark matter plays a crucial role in the formation and evolution of galaxies. Its gravitational pull helps hold galaxies together, prevents them from flying apart, and shapes the large-scale structure of the universe.

- The distribution of dark matter is thought to be clumpy, forming halos around galaxies and galaxy clusters. These halos provide the gravitational scaffolding for the formation of galaxies and help explain the observed rotational velocities of stars within galaxies.

- Dark matter interacts with ordinary matter only through gravity, making it difficult to study in laboratory settings. Scientists rely on a variety of observational techniques, including gravitational lensing and the study of cosmic microwave background radiation, to probe the effects of dark matter on the universe.

- The search for dark matter is ongoing, employing various approaches such as underground experiments, particle accelerators, and astronomical observations. Scientists are continually refining their understanding of dark matter's properties and exploring new avenues for its detection.

- Solving the mystery of dark matter could have profound implications, not just for astrophysics but also for our understanding of particle physics, the early universe, and the fundamental nature of reality itself.

B. Dark Energy and Its Role in the Accelerating Expansion of the Universe

- Dark energy is an even more enigmatic concept than dark matter. It is a hypothetical form of energy that is believed to permeate all of space and is responsible for the accelerating expansion of the universe.

- The existence of dark energy was inferred from observations of distant supernovae in the late 1990s. These observations indicated that the expansion of the universe is not slowing down as expected but instead is accelerating.

- Dark energy is thought to exhibit negative pressure, which produces a repulsive gravitational effect, counteracting the attractive force of gravity between matter and causing the expansion of the universe to accelerate.

- The nature of dark energy is still largely unknown. One possibility is that it arises from the energy associated with empty space, known as vacuum energy or the cosmological constant. Other theories propose the existence of exotic fields or modifications to the laws of gravity.

- Dark energy is believed to be the dominant component of the universe, comprising roughly 70% of its total energy density. Ordinary matter and dark matter make up the remaining 5% and 25%, respectively.

- The accelerated expansion driven by dark energy has far-reaching consequences for the future of the universe. It implies that galaxies outside our local group will eventually move beyond our cosmic horizon, making them unobservable in the distant future.

- The nature and properties of dark energy are actively investigated through various observational techniques, including measuring the cosmic microwave background radiation, large-scale galaxy surveys, and studying the distribution of galaxies and galaxy clusters.

- Understanding dark energy is essential for determining the ultimate fate of the universe. Depending on its properties, the universe could experience a "Big Freeze," where expansion continues indefinitely, or a "Big Rip," where dark energy becomes stronger, tearing apart structures and even subatomic particles.

- The discovery of dark energy revolutionized cosmology and earned the 2011 Nobel Prize in Physics for the astronomers who made the initial observations. It opened up new avenues of research and prompted a deeper investigation into the fundamental forces and constituents of the universe.

- Further exploration of dark energy is a primary focus of current and future space missions and observatories, such as the Euclid mission and the Large Synoptic Survey Telescope (LSST). These efforts aim to refine our understanding of dark energy, its origin, and its role in the evolution of the cosmos.

C. Current Research and Theories Surrounding These Cosmic Enigmas

- Numerous experiments are underway to directly detect dark matter particles. These experiments include deep underground detectors, such as the Large Underground Xenon (LUX) and XENON1T experiments, which aim to observe rare interactions between dark matter and ordinary matter.

- Particle accelerators, such as the Large Hadron Collider (LHC), are also being used to search for dark matter particles. These experiments attempt to recreate the conditions present in the early universe, where dark matter could have been produced.

- The study of galactic dynamics and gravitational lensing provides important insights into the distribution of dark matter. Observations of galaxy rotation curves and the bending of light around massive objects help constrain theories about the nature of dark matter.

- Various theoretical models propose alternative explanations for dark matter. Some theories suggest modifications to the laws of gravity, such as Modified Newtonian Dynamics (MOND), while others invoke extra dimensions or modifications to general relativity.

- The concept of dark energy has led to the development of new theoretical frameworks, such as quintessence and the Chaplygin gas model. These models attempt to explain the nature of dark energy and its behaviour over cosmic time.

- Cosmological surveys, such as the Sloan Digital Sky Survey (SDSS) and the Dark Energy Survey (DES), collect vast amounts of data to map the large-scale structure of the universe and investigate the properties of dark matter and dark energy.

- The study of cosmic microwave background radiation, the afterglow of the Big Bang, provides crucial insights into the early universe and the composition of matter and energy. Precise measurements of the cosmic microwave background are obtained by experiments like the Planck satellite and the upcoming Simons Observatory.

- The use of gravitational wave detectors, such as the Laser Interferometer Gravitational-Wave Observatory (LIGO) and the future Laser Interferometer Space Antenna (LISA), could offer new avenues for studying the properties of dark matter and probing the early universe.

- Synergies between different observational techniques, such as combining data from galaxy surveys, cosmic microwave background experiments, and gravitational wave detections, hold the potential to provide a more comprehensive understanding of dark matter, dark energy, and the fundamental nature of the universe.

- Ongoing research and collaborations between astrophysicists, cosmologists, particle physicists, and theorists aim to unravel the mysteries of dark matter and dark energy. Advancements in technology, data analysis techniques, and theoretical frameworks continue to push the boundaries of our knowledge and bring us closer to unlocking the secrets of these cosmic enigmas.

V. Chapter 5: Exoplanets and the Search for Life

A. The Discovery of Exoplanets and Their Potential for Habitability

- The first confirmed detection of an exoplanet orbiting a Sun-like star was announced in 1995. Since then, thousands of exoplanets have been discovered, revolutionizing our understanding of planetary systems.

- The Kepler Space Telescope launched in 2009, played a crucial role in exoplanet discoveries. It detected planets by observing the tiny dips in a star's brightness when a planet passes in front of it, known as the transit method.

- Exoplanets come in a wide range of sizes and compositions. They can be rocky, like Earth, or gas giants, similar to Jupiter or Neptune. Some exoplanets fall into a category called "super-Earths," which are larger than Earth but smaller than gas giants.

- The "hot Jupiters" are a class of exoplanets that orbit very close to their host stars, resulting in scorching temperatures. These planets challenged conventional theories of planetary formation.

- The discovery of exoplanets in the habitable zone, also known as the Goldilocks zone, where conditions may allow liquid water to exist, sparked great interest in the potential for finding life beyond Earth.

- The TRAPPIST-1 system, discovered in 2017, gained significant attention as it contains seven Earth-sized planets, three of which are located in the habitable zone. This system represents an exciting target for further study.

- The transit method and the radial velocity method, which detects the gravitational pull of an exoplanet on its host star, are the primary techniques used to detect and characterize exoplanets. Other methods, such as direct imaging and gravitational microlensing, have also contributed to exoplanet discoveries.

- The discovery of exoplanets has challenged the notion that our solar system is unique. It suggests that planetary systems are common in the universe and that there may be billions of potentially habitable planets in our galaxy alone.

- The study of exoplanet atmospheres is crucial in assessing their potential habitability. By analyzing the composition of an exoplanet's atmosphere, scientists can search for signs of life or habitable conditions.

- The James Webb Space Telescope, set to launch in 2021, will greatly enhance our ability to study exoplanets. It will provide detailed observations of exoplanet atmospheres and potentially detect biosignatures, such as the presence of oxygen or methane.

B. The Characteristics of Potentially Habitable Exoplanets

- The habitable zone around a star refers to the range of distances where a planet could have surface temperatures suitable for liquid water to exist. It depends on the star's size, temperature, and brightness.

- The presence of liquid water is considered a crucial ingredient for life as we know it. While it is a key factor in assessing a planet's habitability, it does not guarantee the presence of life.

- Exoplanets in the habitable zone can have a wide range of atmospheric compositions, including greenhouse gases that could lead to extreme heat or cold. Understanding the balance of gases is essential in determining a planet's habitability.

- The mass and composition of an exoplanet also play significant roles in its habitability. Rocky planets with a similar mass to Earth are more likely to have solid surfaces, which can potentially support liquid water and allow for the development of life.

- The presence of an exoplanet's magnetic field is crucial for protecting its atmosphere and surface from harmful stellar radiation. A strong magnetic field, like Earth's, can help sustain a habitable environment.

- The rotation rate of an exoplanet affects its climate and the distribution of heat across its surface. A moderate rotation rate is considered favourable for maintaining a stable climate and potentially supporting life.

- The presence of stable plate tectonics on an exoplanet is thought to be beneficial for habitability. Plate tectonics helps regulate the carbon cycle, control the climate, and provide nutrient cycling through volcanic activity.

- The distance between a planet's star and its moon can influence its habitability. A moon's gravitational interactions with its planet can generate tidal heating, which can provide additional energy and potentially support life.

- The presence of an atmosphere is crucial for a planet's habitability. It can regulate temperature, distribute heat, and provide protection from harmful radiation. The composition and stability of the atmosphere play vital roles in supporting life.

- The potential for habitable exomoons, orbiting gas giants in the habitable zone, has also been considered. These exomoons could offer a unique environment where life could thrive, benefiting from the planet's protection and potential energy sources.

C. The Ongoing Search for Extraterrestrial Life and the Implications for Humanity

- The search for extraterrestrial life encompasses a broad range of approaches, including the exploration of Mars, the study of microbial life in extreme environments on Earth, and the analysis of exoplanet atmospheres for biosignatures.

- The discovery of even microbial life beyond Earth would have profound implications for our understanding of life's origin and its prevalence in the universe.

- The Drake Equation, formulated by astronomer Frank Drake, attempts to estimate the number of technologically advanced civilizations in our galaxy. While the equation involves many uncertainties, it highlights the potential for intelligent life elsewhere.

- The study of extremophiles, organisms that thrive in extreme conditions on Earth, has expanded our understanding of where and how life can exist. It suggests that life might be resilient and adaptable, even in challenging environments.

- NASA's Mars exploration missions, such as the Mars rovers and the upcoming Mars Sample Return mission, aim to search for evidence of past or present life on the Red Planet. The study of Mars provides insights into the potential habitability of other planets.

- The study of extremophiles on Earth's ocean floors, hydrothermal vents, and icy moons, such as Enceladus and Europa, has raised the possibility of finding life in similar environments within our solar system.

- The search for technosignatures, such as radio signals or other technological artefacts, is an active area of research. Initiatives like the Breakthrough Listen project scan the skies for signals from intelligent civilizations.

- The potential discovery of extraterrestrial life, even in microbial form, would raise profound philosophical and societal questions, including our place in the universe, the nature of life, and the ethical implications of contact.

- The detection of biosignatures, such as atmospheric gases indicative of life, on an exoplanet would represent a significant breakthrough. It could inform future missions and the development of telescopes capable of characterizing distant worlds.

- The ongoing search for extraterrestrial life fuels our curiosity and drives technological advancements. It inspires collaboration between scientists, encourages public engagement, and stimulates discussions about our place in the cosmos.

VI. Chapter 6: Cosmic Events and Phenomena

A. Eclipses, Meteor Showers, and Other Celestial Events

- A solar eclipse occurs when the Moon passes between the Sun and Earth, casting a shadow on Earth's surface. It creates a breathtaking sight as the Moon temporarily blocks the Sun, revealing its outer atmosphere, known as the corona.

- Lunar eclipses occur when Earth comes between the Sun and the Moon, casting its shadow on the Moon's surface. During a total lunar eclipse, the Moon can appear reddish due to Earth's atmosphere bending sunlight around the planet and onto the Moon.

- Meteor showers occur when Earth passes through the debris left behind by comets or asteroids. These tiny particles, known as meteoroids, burn up upon entering Earth's atmosphere, creating streaks of light called meteors. Examples of famous meteor showers include the Perseids and the Geminids.

- The auroras, also known as the Northern and Southern Lights, are breathtaking displays of light in the Earth's polar regions. They are caused by charged particles from the Sun colliding with molecules in Earth's atmosphere, creating colourful and dynamic patterns in the sky.

- Transits of Mercury and Venus are rare celestial events where these planets pass directly between the Sun and Earth, appearing as small black dots moving across the Sun's surface. They provide valuable opportunities for scientific observations and measurements.

- Planetary conjunctions occur when two or more planets appear close together in the sky. These events create visually striking configurations and offer opportunities for stargazers to witness unique celestial alignments.

- Comet sightings are thrilling events as these icy bodies from the outer reaches of the solar system journey through the inner solar system. Comets develop glowing tails as they approach the Sun, making them visible to the naked eye.

- The annual solstices mark the points in Earth's orbit around the Sun when the tilt of Earth's axis is most inclined toward or away from the Sun. They correspond to the longest and shortest days of the year, respectively, and are significant events in various cultures.

- The equinoxes occur when the plane of Earth's equator passes through the centre of the Sun, resulting in equal durations of day and night across the globe. These events signal the changing of seasons and are important markers in astronomical calendars.

- Occultations happen when one celestial object appears to pass in front of another, temporarily blocking it from view. These events allow astronomers to study the occulted object and gather valuable data about its characteristics.

B. Gamma-Ray Bursts and Their Energetic Impact

- Gamma-ray bursts (GRBs) are the most energetic explosions in the universe, releasing enormous amounts of gamma-ray radiation. They can outshine entire galaxies for brief periods.

- GRBs are classified into two categories: long-duration bursts and short-duration bursts. Long-duration bursts, lasting a few seconds to minutes, are associated with the collapse of massive stars. Short-duration bursts, lasting less than two seconds, are thought to arise from the merger of compact objects like neutron stars or black holes.

- The exact mechanisms that produce GRBs are still under investigation, but one leading theory suggests that they result from the formation of a black hole or a highly magnetized neutron star during a supernova event.

- GRBs emit radiation across the electromagnetic spectrum, including gamma rays, X-rays, and sometimes visible light. Observatories like NASA's Swift satellite and the Fermi Gamma-ray Space Telescope have greatly contributed to our understanding of GRBs.

- The afterglow of a GRB, which occurs at longer wavelengths than the initial burst, can last from hours to weeks. Studying the afterglow provides valuable information about the explosion and its environment.

- GRBs can have a profound impact on their host galaxies. The intense radiation from a GRB can ionize the gas in its vicinity, trigger star formation, and even affect the chemistry of surrounding regions.

- The discovery of GRBs in the late 1960s was initially classified as a military secret due to concerns about potential nuclear detonations. It was later declassified, allowing scientists to study these cosmic phenomena.

- GRBs played a significant role in confirming the prediction of the existence of gamma-ray bursts by astrophysicist Yakov Zel'dovich and mathematician Yakov G. Perelemov in 1972. Their theoretical work laid the foundation for understanding the energetic nature of these events.

- GRBs are detected from sources across the universe, indicating that they occurred billions of years ago due to the finite speed of light. Studying these distant bursts provides insights into the early universe.

- The study of GRBs has led to advancements in understanding high-energy astrophysics, the physics of black holes and neutron stars, and the processes that shape galaxies. It continues to be an active area of research, combining observations from ground-based telescopes, space-based missions, and theoretical modelling.

C. Gravitational Waves: Ripples in the Fabric of Spacetime

- Gravitational waves are ripples in the fabric of spacetime caused by the acceleration of massive objects. They were predicted by Albert Einstein's general theory of relativity in 1915.

- The first direct detection of gravitational waves occurred in 2015 by the Laser Interferometer Gravitational-Wave Observatory (LIGO). This groundbreaking discovery confirmed Einstein's theory and opened a new window to observe the universe.

- Gravitational waves carry information about their source and the events that generated them. By studying the properties of gravitational waves, scientists can gain insights into phenomena such as the mergers of black holes and neutron stars.

- LIGO and its European counterpart, Virgo, are ground-based interferometers designed to detect gravitational waves. They consist of two or more arms several kilometres long, with laser beams travelling back and forth, precisely measuring any length changes caused by passing gravitational waves.

- Gravitational waves are produced by a variety of astrophysical events, including the collision of black holes, the merger of neutron stars, and the early moments of the universe during the Big Bang. Each event leaves a unique imprint on the gravitational wave signal.

- Gravitational waves are capable of traversing vast cosmic distances without being significantly affected by intervening matter. This allows scientists to observe events that may be obscured by gas, dust, or other cosmic structures.

- The detection of gravitational waves has enabled the direct measurement of properties such as the masses and spins of black holes and neutron stars, providing crucial insights into their formation and evolution.

- The study of gravitational waves provides a new way to test the predictions of general relativity and explore the nature of gravity itself. It opens avenues for investigating physics in extreme environments that are difficult to replicate in laboratories.

- The merger of two neutron stars detected through gravitational waves in 2017 also produced a gamma-ray burst, confirming the long-suspected connection between these two cosmic phenomena.

- The field of gravitational wave astronomy is rapidly advancing, with ongoing efforts to improve detectors, increase the number of observing facilities, and expand international collaborations. The future holds the promise of more detections, unveiling new aspects of the universe and providing insights into unresolved astrophysical mysteries.

VII. Chapter 7: Space Exploration and Discoveries

A. Milestones in Space Exploration: From the Moon Landing to Mars Missions

- The Apollo 11 mission in 1969 marked a historic milestone as astronauts Neil Armstrong and Buzz Aldrin became the first humans to walk on the Moon's surface.

- The Voyager 1 and Voyager 2 spacecraft, launched in 1977, provided humanity's first close-up views of the outer planets of our solar system, including Jupiter, Saturn, Uranus, and Neptune.

- The Hubble Space Telescope, launched in 1990, revolutionized our understanding of the universe by capturing breathtaking images and making groundbreaking observations of distant galaxies, nebulae, and other celestial objects.

- The Mars Pathfinder mission in 1997 successfully landed the Sojourner rover on Mars, marking the first successful deployment of a robotic rover on the red planet's surface.

- The Mars rovers Spirit and Opportunity, launched in 2003, greatly expanded our knowledge of Mars by studying its geology, searching for signs of past water, and operating well beyond their expected mission lifetimes.

- The Cassini-Huygens mission, launched in 1997, provided detailed insights into the Saturn system, including its rings, moons, and the landing of the Huygens probe on Saturn's largest moon, Titan.

- The successful landing of NASA's Mars Science Laboratory mission in 2012 introduced the Curiosity rover, equipped with advanced scientific instruments to investigate Mars' habitability and search for evidence of past or present microbial life.

- The New Horizons mission, launched in 2006, conducted a historic flyby of Pluto and its moons in 2015, capturing the first detailed images of this distant dwarf planet and revealing surprising features on its surface.

- The International Space Station (ISS), a collaborative effort among multiple countries, has been continuously inhabited since 2000 and serves as a valuable platform for scientific research, technological development, and international cooperation in space.

- The Artemis program, led by NASA, aims to return humans to the Moon by 2024, to establish sustainable lunar exploration and pave the way for future crewed missions to Mars.

B. Robotic Probes and Their Contributions to Our Understanding of the Universe

- The Pioneer 10 and 11 missions, launched in the early 1970s, provided the first close-up images of Jupiter and Saturn, respectively. They also paved the way for future interstellar missions by carrying plaques with information about Earth in case they were ever encountered by extraterrestrial civilizations.

- The Viking missions, launched in 1975, were the first to land spacecraft on Mars. They conducted experiments to search for signs of life and provided valuable data on the Martian environment.

- The Galileo spacecraft, launched in 1989, extensively studied Jupiter and its moons, discovering evidence of subsurface oceans on Jupiter's moon Europa and capturing stunning images of the Jovian system.

- The Mars Reconnaissance Orbiter (MRO), launched in 2005, has been instrumental in providing high-resolution images of Mars' surface, monitoring its weather patterns, studying its atmosphere, and identifying potential landing sites for future missions.

- The Kepler space telescope, launched in 2009, revolutionized exoplanet discoveries by identifying thousands of planets orbiting distant stars, significantly expanding our understanding of planetary systems beyond our own.

- The Dawn spacecraft, launched in 2007, visited the asteroid Vesta and the dwarf planet Ceres, providing detailed insights into their geology, composition, and history. It revealed diverse landscapes and even detected signs of potential subsurface oceans.

- The Juno spacecraft, launched in 2011, has been studying Jupiter since it arrived in 2016, providing unprecedented data on the planet's atmosphere, magnetic field, and internal structure. It aims to unravel the mysteries of Jupiter's formation and evolution.

- The Rosetta mission, launched in 2004, successfully placed a lander named Philae on the surface of a comet called 67P/Churyumov-Gerasimenko in 2014. It provided valuable data on cometary composition and structure, shedding light on the early solar system.

- The Parker Solar Probe, launched in 2018, is on a mission to study the Sun up close, entering its outer atmosphere, known as the corona, and collecting data to help scientists better understand solar activity and its effects on space weather.

- The James Webb Space Telescope (JWST), set to launch in 2021, promises to be the most powerful space telescope ever built. It will observe the universe in infrared wavelengths, providing unprecedented views of distant galaxies, star formation, and exoplanets.

C. The Future of Space Exploration and the Prospects for Interstellar Travel

- The development of reusable rocket systems, such as SpaceX's Falcon 9 and Falcon Heavy, and Blue Origin's New Shepard and New Glenn, is revolutionizing space exploration by significantly reducing the cost of launching payloads and paving the way for more ambitious missions.

- NASA's Artemis program aims to establish a sustainable human presence on the Moon, with the long-term goal of using it as a stepping stone for crewed missions to Mars and deeper into the solar system.

- The concept of asteroid mining is gaining attention as companies explore the potential for extracting valuable resources from asteroids, such as water, metals, and rare minerals. This could provide a sustainable source of materials for future space missions and benefit industries on Earth.

- The exploration of icy moons, such as Europa (a moon of Jupiter) and Enceladus (a moon of Saturn), is of great interest due to their potential subsurface oceans and the possibility of finding extraterrestrial life or habitable environments.

- The Breakthrough Starshot initiative aims to send small, lightweight spacecraft to nearby star systems using ultra-powerful lasers for propulsion. The goal is to achieve interstellar travel within a human lifetime and explore other star systems up close.

- The concept of establishing permanent human colonies on Mars is actively being studied by space agencies and private companies. Initiatives like SpaceX's Starship envision sending large groups of people to the red planet to establish self-sustaining habitats.

- Advances in propulsion systems, such as ion propulsion and nuclear propulsion, are being explored to enable faster and more efficient space travel, reducing travel times and opening up new possibilities for deep space exploration.

- The study of exoplanets is advancing rapidly, to identify potentially habitable worlds and search for signs of life beyond Earth. Future missions, such as NASA's James Webb Space Telescope and the European Space Agency's PLATO mission, will provide more detailed observations of exoplanet atmospheres and compositions.

- The concept of space tourism is becoming a reality, with companies like SpaceX and Blue Origin planning to offer commercial trips to space for private individuals. This could open up space exploration to a broader audience and generate new opportunities for research and development.

- International collaborations, such as the International Lunar Gateway and the European Space Agency's involvement in Mars missions, highlight the cooperative nature of future space exploration. These partnerships enable shared resources, expertise, and cost-sharing, making ambitious missions more achievable.

VII. Chapter 8: The Human Connection to the Universe

A. Cultural and Historical Perspectives on the Cosmos

- Ancient civilizations, such as the Egyptians, Greeks, and Mayans, developed elaborate cosmologies that integrated celestial observations into their religious and cultural beliefs.

- The Babylonians, one of the earliest known civilizations, were skilled astronomers and created one of the first recorded astronomical diaries, known as the "Enuma Anu Enlil."

- The Islamic Golden Age (8th to 14th centuries) saw significant advancements in astronomy, with scholars like Al-Battani and Al-Khwarizmi making notable contributions to celestial observations and mathematical calculations.

- In ancient China, astronomy held great importance, and records of celestial phenomena were meticulously kept. Chinese astronomers made groundbreaking observations, such as the recording of supernovae and the discovery of the precession of Earth's axis.

- The Renaissance period witnessed a resurgence of interest in astronomy, fueled by advancements in mathematics and the development of telescopes. Scientists like Nicolaus Copernicus, Johannes Kepler, and Galileo Galilei revolutionized our understanding of the cosmos.

- The invention of the printing press in the 15th century played a crucial role in disseminating astronomical knowledge and expanding public awareness of celestial phenomena.

- The development of accurate star catalogues, such as Tycho Brahe's "Uranometria" and Johann Bayer's "Uranometria," facilitated the identification and classification of stars, constellations, and celestial objects.

- The Space Age, beginning with the launch of Sputnik 1 in 1957, sparked a new era of exploration and heightened public fascination with space, leading to significant advancements in astronomy, astrophysics, and space technology.

- The establishment of organizations like the International Astronomical Union (IAU) and the formation of global collaborations, such as the Square Kilometre Array (SKA) project, have facilitated international cooperation in astronomical research and discovery.

- Indigenous cultures worldwide possess rich celestial knowledge, with many indigenous communities incorporating astronomy into their cultural practices, folklore, and navigation techniques, demonstrating the deep connection between humanity and the cosmos.

B. The Role of Astronomy in Shaping Our Understanding of the World

- Astronomy has played a pivotal role in developing our understanding of the fundamental laws of physics and the nature of the universe itself.

- Through astronomical observations, scientists have made groundbreaking discoveries, such as the laws of planetary motion formulated by Kepler and Newton's laws of gravitation.

- The study of stellar evolution and nuclear fusion in stars has provided insights into the origin of chemical elements and the processes that drive the energy production in the universe.

- Astronomy has contributed to our understanding of Earth's place in the cosmos, highlighting our planet's uniqueness and the delicate balance necessary to support life.

- The discovery and study of exoplanets have expanded our understanding of planetary systems and the potential prevalence of habitable worlds beyond our solar system.

- Astronomical observations of cosmic microwave background radiation have provided crucial evidence for the Big Bang theory, supporting the idea that the universe originated from a hot, dense state.

- The study of cosmic rays and high-energy particles has deepened our knowledge of particle physics and the interactions of matter and energy in extreme environments.
- Astronomy has advanced our understanding of the origins and evolution of galaxies, shedding light on the formation of structures in the universe and the distribution of dark matter.
- The exploration of cosmic phenomena, such as supernovae, gamma-ray bursts, and black holes, has expanded our knowledge of extreme astrophysical processes and the behavior of matter under extreme conditions.
- Astronomy has fostered interdisciplinary collaborations, allowing scientists to combine expertise from various fields, including physics, chemistry, biology, and computer science, to tackle complex questions and advance our understanding of the universe.

C. Reflections on Our Place in the Vastness of the Universe

- The vastness of the universe, with its billions of galaxies and trillions of stars, humbles us and invites contemplation of our place in the cosmic tapestry.
- Astronomical observations have shown that our planet is just a tiny speck in the vastness of the universe, emphasizing the need for humility and a broader perspective.
- The concept of the "pale blue dot," coined by Carl Sagan, underscores the fragility and interconnectedness of Earth and reminds us of the importance of preserving and protecting our planet.
- Astronomy fosters a sense of wonder and curiosity about the universe, prompting philosophical and existential reflections on our existence and purpose.
- The study of cosmology and the discovery of the accelerating expansion of the universe raise questions about the ultimate fate of the cosmos and our place within it.
- The search for extraterrestrial life and the possibility of other intelligent civilizations provoke contemplation of our uniqueness or potential cosmic companions.
- The study of dark matter and dark energy, which constitute the majority of the universe's content, underscores our limited understanding of the cosmos and the mysteries that remain to be unraveled.
- Astronomy encourages a sense of unity among humanity, as we collectively explore and seek to comprehend the vastness and complexity of the universe.

- The cosmic perspective provided by astronomy inspires a sense of awe, fostering a deeper appreciation for the beauty, complexity, and interconnectedness of the natural world.

- The exploration of space and the search for knowledge beyond our planet reflect our innate curiosity, drive for exploration, and the human desire to expand our horizons, both literally and metaphorically.

IX. Chapter 9: Bonus Facts about Space

- Time dilation: As you approach the speed of light, time slows down relative to an observer at rest. This means that astronauts who spend significant time in space, particularly in high-speed travel, age slightly slower compared to people on Earth.

- White holes: Theoretically, white holes are the reverse of black holes. While black holes pull in matter and light, white holes are hypothesized to emit matter and light but do not allow anything to enter. However, white holes have not been observed directly.

- Gravitational lensing: Massive objects in space, such as galaxies and galaxy clusters, can bend and distort light passing through their gravitational fields, creating gravitational lensing. This phenomenon allows astronomers to observe distant galaxies that would otherwise be too faint to detect.

- The Pillars of Creation: Located in the Eagle Nebula, the Pillars of Creation are massive columns of gas and dust where new stars are forming. The iconic image captured by the Hubble Space Telescope showcases the incredible beauty and complexity of the universe.

- Space is not completely empty: Although space is often thought of as empty, it contains trace amounts of gas and dust spread throughout. In certain regions, these particles can come together to form dense interstellar clouds, giving birth to stars and planetary systems.

- The Great Attractor: The Great Attractor is a gravitational anomaly located in the direction of the Centaurus and Hydra constellations. It exerts a gravitational force that pulls galaxies, including our Milky Way, towards it. The exact nature of the Great Attractor is still not fully understood.

- Vacuum energy: According to quantum field theory, even empty space is not truly empty. It is filled with virtual particles that spontaneously pop in and out of existence. These fluctuations contribute to what is known as vacuum energy, which may play a role in the expansion of the universe.

- The Largest Known Structure: The Hercules-Corona Borealis Great Wall is considered the largest known structure in the universe. It is a massive filament of galaxies spanning about 10 billion light-years across.

- Galactic cannibalism: Galaxies can merge and collide with each other, resulting in a process known as galactic cannibalism. Larger galaxies can consume smaller ones, incorporating their stars, gas, and dust. Our Milky Way is expected to collide with the neighbouring Andromeda Galaxy in about 4 billion years.

- The universe is expanding faster than expected: Recent observations have revealed that the universe is expanding at an accelerating rate. This discovery led to the proposal of dark energy, a mysterious force driving the accelerated expansion and comprising a significant portion of the universe's total energy.

THE END

www.ingramcontent.com/pod-product-compliance
Lightning Source LLC
Chambersburg PA
CBHW050326220526
45465CB00005B/2145